Ra
I.
Pocket Guide

RFI Characterization, Location
Techniques, Tools and Remediation
Methods, with Key Equations and Data

Kenneth Wyatt, WA6TTY
Michael Gruber, W1MG

Special thanks to our technical reviewers and RFI experts:
David Eckhardt, EMC consultant, W0LEV; Ed Hare, ARRL lab
manager, W1RFI; Kit Haskins, broadcast engineer, KA0WUC;
Jon Sprague, FCC engineer-retired, WB7UIA; and Robert
Witte, VP-R&D Keysight Technologies, K0NR.

SCITECH
PUBLISHING
an imprint of the IET

Published by SciTech Publishing, an imprint of the IET
www.scitechpub.com
www.theiet.org

ISBN 978-1-61353-219-5 (spiral bound)
ISBN 978-1-61353-220-1 (PDF)

Typeset in India by MPS Ltd
Printed in the USA by Docusource (Raleigh, NC)

Contents

■ Introduction

Thanks for purchasing the *RFI Pocket Guide*. The purpose of this guide to help you identify, locate and resolve radio frequency interference (RFI). It includes some basic theory and measurement techniques and there are a number of handy references, tables, and equations that you may find useful. The focus is to assist both amateur radio operators, as well as commercial broadcast and communications engineers, in resolving a variety of common interference issues.

As you read through this guide, you'll note two primary interference locating techniques: use of receivers and use of spectrum analyzers. For many amateur radio operators, simply using a receiver to track down the interference source will be sufficient. However, for more complex interference sources, the spectrum analyzer may be the tool of choice. Keep this in mind as you use the information in this guide. Good luck!

■ EMC/RFI Fundamentals

What is EMC?

Electromagnetic Compatibility (EMC) is achieved when:

- Emissions from electronic products do not interfere with their environment.
- The environment does not upset the operation of electronic products; that is, they are immune.
- Electronic products do not interfere with themselves. (signal integrity)

In reviewing the various ways signals can be propagated within and between systems, we see that energy is transferred from source to receiver (victim) via some coupling path (Figure 1).

Conducted Emissions (CE) – Radio frequency energy that is generated by an electronic device but emanates from it via other conductors connected to it, such as an AC power cord. Although the RF can then be conducted directly to the victim, the typical path also includes radiation from these conductors.

Conducted emissions from the AC power connection are regulated by the U.S. Federal Communications Commission (FCC) for most electronic products because the energy can then be conducted to and radiated by the associated house wiring. In some cases, it can then be conducted to and radiated by the service entrance and utility hardware. This larger network of power line conductors can then radiate the energy more

---

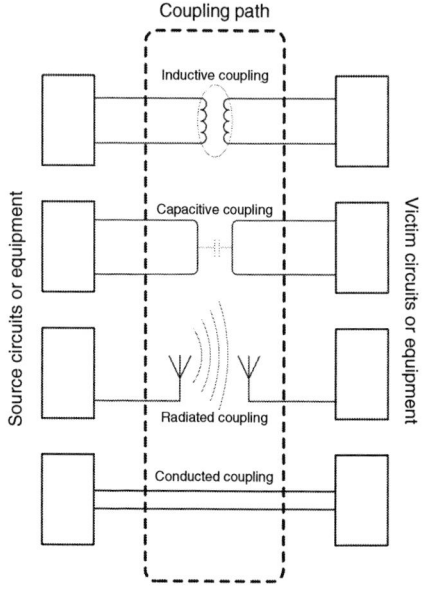

FIGURE 1 Key RFI–EMC interaction relationships.

efficiently than the source device could by itself, especially at lower frequencies. For this reason, the FCC only imposes conducted emissions limits below 30 MHz in the United States.

Radiated Emissions (RE) – Radio frequency energy that is generated by an electronic device and emanates from it via radiation. In the United States, the FCC only imposes radiated emissions limits from electronic devices above 30 MHz. The field strength of this energy is measured with cables and wires connected to the device and located in the same manner as the user would install them, or manipulated within the range of likely arrangements, depending on the device.

Removing any of these three – Source, Coupling Path, or Victim – will eliminate EMC problems.

What is RFI?

Radio Frequency Interference is caused by:

- The disruption of an electronic device or system due to external electromagnetic emissions at radio frequencies (usually a few kHz to a few GHz). Also see EMI in the list of definitions at the end of the Guide.
- Electronic products, other transmitters, or RF energy sources that interfere with radio reception.

Very often, RFI issues are frequency related (digital harmonics, switching power supply "noise", or other transmitters) and are best

---

identified using a spectrum analyzer or, for power line noise, simply a portable amplitude modulation (AM) broadcast or high-frequency (HF) receiver, depending on the interfering signal frequency. While power line interference is often best identified using "signature analysis" (a time-domain technique used by professional RFI investigators), a simple AM broadcast or, or even better, a portable HF receiver can also help identify and locate the source in many cases.

A common interference issue to radio reception at HF, and lower frequencies, involves conducted emissions from a consumer product, then radiation from the conductors to the antenna of the receiver. Radiated emissions from consumer devices also tend to be more problematic at very high frequency (VHF) and higher frequencies.

Another common source of radio interference is power line noise, which is typically caused by arcing on commercial power lines or related hardware. Sometimes called "gap noise" in the power industry, a typical path usually includes both conduction and radiation. Although less common, the path in a power line noise case can also involve induction. Contrary to common belief, corona discharge is rarely the cause of a power line noise problem.

In most cases, the solution to a conducted emissions problem from a consumer device involves isolating (or filtering) radiating cables using ferrite chokes or discrete filters at the source device – see the section on "Hidden Antennas" below. Correcting a power line noise problem, however, typically requires fixing the defect that is causing it. This is a job for the power company!

Digital Signal Spectra

Most internally generated interference originates from fast-switching digital signals or clock generators.

Figure 2 shows a trapezoidal waveform that represents the output of the clock for digital circuits. The faster the rise time, τ_r, and fall time, τ_f, (typically about the same), the higher the interfering harmonics in frequency. For example, a clock frequency of 10 MHz will also produce

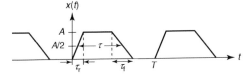

FIGURE 2 A typical trapezoidal digital waveform.

---

higher-order harmonics every 10 MHz (example, 20, 30, 40, 50 ... MHz). With today's 1 nanosecond rise times (or less) it's not unusual to generate clock or switching harmonics in the hundreds of MHz. Note that pure sine waves do not have these very fast rise times and do not contain any harmonic energy.

Transmitter Harmonics

Because all transmitters include non-linearities in their circuitry, they tend to produce harmonics from the fundamental frequency. The FCC requires harmonics and other spurious emissions to be below certain limits.

There are several specifications on harmonic suppression. This is dependent upon where in the RF spectrum you are involved. For example, Part 73 (the FCC rules involving the broadcast service) has a point around 5 kW output power that requires the harmonic level (in dBc) to be $43+ \log$(power in watts) or -80 dBc, whichever is less. For Part 90 (LMR/Paging/etc.), it's -80 dBc. Amateur radio is a little more relaxed in the requirements of harmonic content. CFR Part 97.307(e) specifies less than 43 dBc below 30 MHz and less than 60 dBc from 30 to 255 MHz.

■ Frequency Versus Wavelength

Most RFI issues occur in the range 9 kHz through 6 GHz. Problems resulting from conducted emissions tend to occur below 30 MHz and radiated emissions tend to occur above 30 MHz, due to the fact that product cables tend to become more efficient radiators above 30 MHz.

The Electromagnetic Spectrum

This is a table of the International Telecommunications Union (ITU) radio bands and the Institute of Electrical and Electronics Engineers (IEEE) radar/microwave bands. Most RFI issues will fall in these bands.

Band	Freq. Range	Wavelength
VLF	3–30 kHz	100,000–10,000 m
LF	30–300 kHz	10,000–1,000 m
MF	0.3–3.0 MHz	1000–100 m
HF	3–30 MHz	100–10 m
VHF	30–300 MHz	10–1 m
UHF	300–3000 MHz	1 m–10 cm
SHF	3–30 GHz	10–1 cm
L	1–2 GHz	30–15 cm
S	2–4 GHz	15–7.5 cm
C	4–8 GHz	7.5–3.75 cm
X	8–12 GHz	3.75–2.5 cm

Band	Freq. Range	Wavelength
K_u	12–18 GHz	2.5–1.67 cm
K	18–27 GHz	1.67–1.11 cm
K_a	27–40 GHz	1.11–0.75 cm

For more detail on the users of the EM Spectrum review, see the chart at:

http://www.ntia.doc.gov/files/ntia/publications/spectrum_wall_chart_aug2011.pdf. Note that this is the spectrum allocation chart for the USA. Other countries have similar allocation charts.

Also, Rohde & Schwarz has a high-resolution and "zoomable" version of the above chart built in to their Interference Hunter iPad or Android app (see References section below).

Frequency Versus Wavelength (free space)

Freq.	Wavelength	1/4 λ	1/2 λ
10 Hz	30,000 km	7,500 km	15,000 km
60 Hz	5,000 km	1,250 km	2500 km
400 Hz	750 km	187.5 km	375 km
1 kHz	300 km	75 km	150 km
10 kHz	30 km	7.5 km	15 km
100 kHz	3 km	750 m	1.5 km
1 MHz	300 m	75 m	150 m
10 MHz	30 m	7.5 m	15 m
100 MHz	3 m	75 cm	150 cm

Freq.	Wavelength	1/4 λ	1/2 λ
300 MHz	100 cm	25 cm	50 cm
500 MHz	60 cm	15 cm	30 cm
1 GHz	30 cm	7.5 cm	15 cm
10 GHz	3 cm	0.75 cm	1.5 cm

Hidden Antennas

An important concept to grasp is the electrical dimension of an electromagnetic radiating structure. EMC engineers often call any radiator of electromagnetic energy an "antenna," whether it is an actual antenna or another radiator, such as a cable or circuit-board trace. This is expressed in terms of *wavelength* (λ).

In a lossless medium (free space)

wavelength $= \lambda = v_o/f$
$v =$ velocity of propagation
$v_o =$ the speed of light
$f =$ frequency (Hz)

In free space

$v = v_o \approx 3 \times 10^8$ m/s (approx. speed of light)

This may also be expressed as 1.86×10^5 miles/s or 3×10^5 km/s.

Easy-to-remember formulas for wavelength in free space:

λ (m) $= 300/f$(MHz) or λ (ft) $= 984/f$(MHz)

Note: For practical conductors, the velocity of propagation is less than in free space:

$$\lambda/2 \text{ (ft)} = 492/f\text{(MHz) in free space,}$$
or approximately

$$\lambda/2 \text{ (ft)} = 468/f \text{ (MHz) for a physical half-wave}$$
dipole antenna

This becomes important when it comes to identifying potential radiating structures – so-called "hidden antennas" – of a consumer or industrial product or system that could be the source of interference. These structures could include:

- Cables (I/O or power)
- Seams/slots in shielded enclosures
- Apertures in enclosures
- Poorly bonded sheet metal (of enclosures)
- Internal interconnect cables
- Peripheral equipment connected to the equipment under test (EUT)

For example, as a cable or slot approaches $1/2$ wavelength (or a multiple) at the frequency of concern, it becomes an efficient transmitting or

---

receiving antenna for interference. Use the previous chart, Frequency Versus Wavelength, for help. The solution might be to install ferrite chokes or filters on cables and seal up slots in enclosure seams.

■ Broadcast Frequency Allocations (U.S.)

AM Broadcast: 540 to 1710 kHz, in 10 kHz steps

FM Broadcast: FCC channel 201 (88.1 MHz) to 300 (107.9 MHz). Channel frequencies incremented every odd tenth of a MHz.

Television (all channels 6 MHz wide):

VHF Band: Channel 2 (54–60 MHz) to 6 (82–88 MHz) and channel 7 (174–180 MHz) to channel 13 (210–216 MHz)

UHF Band: Channel 14 (470–476 MHz) to 36 (602–08 MHz) and channel 38 (614–620 MHz) to channel 51 (692–698 MHz)

Note: Channel 37 (608 MHz to 614 MHz) is reserved for radio astronomy. Channels 14–20 have been assigned for land mobile radio use in some areas.

■ Identifying RFI

Categories of Interference

There are two broad categories of interference.

Narrow Band – this would include continuous wave (CW) or modulated CW signals. Examples would include harmonics from crystal oscillators or other fast rise time digital devices, co-channel transmissions, adjacent-channel transmissions, intermodulation products, etc. On a spectrum analyzer, this would appear to be narrow vertical lines or slightly wider modulated vertical bands associated with specific frequencies. This may sound like a single audio tone in a receiver (Figure 3).

Broadband – this would primarily include switch-mode power supply harmonics, arcing in power lines, wide-band digital communications, or possibly commercial broadcast transmissions, such as military spread spectrum communications, Wi-Fi or digital television. On a spectrum analyzer, this would appear to be broad ranges of signals or an increase in the noise floor (Figure 3). Power line or switch-mode power supplies may sound like buzzing or rasping in a receiver, or a hissing sound.

Types of Interference

The sections below describe the most common types of interference.

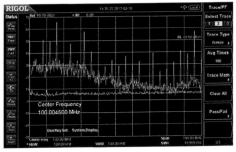

FIGURE 3 An example spectral plot from 9 kHz to 200 MHz of narrow-band harmonics (vertical spikes) riding on top of broadband interference (broad area of increased noise floor).

Co-Channel Interference – More than one transmitter (or digital harmonic) using, or falling into, the same channel.

Adjacent-Channel Interference – A transmitter operating nearby on an adjacent frequency whose energy spills over into the desired channel.

Intermodulation-Based Interference – Occurs when energy from two, or more, transmitters mix together to produce spurious frequencies that land in the desired receive channel. Third-order mixing products are the most common and usually this occurs from nearby transmitters. An example of

---

potential intermodulation might occur in a strong signal area for FM broadcast.

A tougher, but related, problem can be caused by corrosion between two pieces of metal near a transmitter or receiver. This is sometimes called the "rusty-bolt effect." This creates a non-linear junction, which can create and re-radiate harmonics from a single transmitter or intermodulation products from two, or more, transmitters. This can sometimes be found by striking a suspect joint/junction/chain link fence/strained ground braid with a rubber mallet or large insulated screwdriver. This will alter the contact of the corroded junction and will change the RF mixing which can be observed as changes in RFI with the victim receiver or changes of frequency in the spectrum analyzer. Fortunately this is a rare and short-range phenomenon.

An example of the math for intermodulation products is shown in the equation below.

$$n(f_1) \pm m(f_2)$$

where n and m are integers and f_1 and f_2 are interfering signals. If the sum of n and m is odd (2 and 1, 3 and 2, or 3 and 4, etc.), the result is products that have frequencies near the desired signal. If the sum of n and m is 3, those are third-order intermodulation products. The higher the order (3 and 2, for example), the smaller the distortion products in amplitude, so the main concern is with third-order products, for example.

There is also an issue with receiver intermodulation products, especially those used in repeater sites or cellular sites that lack the proper selective filtering "pre-selector stage" that minimizes the extraneous RF from entering the RF pre-amplifier stage. This is pretty common with trunked sites (land mobile radio (LMR), cellular) with multiple receivers on a receiver distribution amplifier. This may be fixed by adding attenuation pads in receiver or very narrow bandpass filters.

Fundamental Receiver Overload – In this case a strong, but spectrally clean, transmitter can simply overload the receiver front-end or other circuitry, causing interference or even suppression (or masking) of the normal received signal. A common example is paging transmitters interfering with amateur VHF receivers. Other types of electronic device can also suffer from fundamental overload, such as audio amplifiers, alarm systems, etc.

Power Line Noise – This is a relatively common broadband interference problem that is typically

---

caused by arcing on electric power lines and associated utility hardware. Contrary to common misconception, it is rarely caused by corona discharge. It sounds like a harsh raspy buzz in an AM receiver. The interference can extend from very low frequencies below the AM broadcast band, and depending on proximity to the source, into the HF spectrum. If close enough to the source, it can extend through VHF and up into the ultra-high frequency (UHF) spectrum and beyond.

Consumer Devices – Switching-mode power supplies used for consumer products are a very common source of interference. Lighting devices, such as the newer LED-based lights, are another. Plasma TVs, electric fences, invisible dog fences, HVAC equipment, and Wi-Fi routers are also common sources. Under the FCC rules, devices capable of causing radio interference are broken down into a variety of categories, depending on the type of device:

FCC Part 15 Devices

- Incidental radiators – A device that generates RF energy during its operation, but is not specifically designed to do so, such as motors, dimmer controls, or light switches. Power lines and related hardware are a common source of interference from an incidental radiator.

- Unintentional radiators – A device that intentionally generates RF energy for use within the device, but is not intended to radiate this energy. For example, this would include clock generators or a local oscillator in a radio receiver.
- Intentional radiators – A device that intentionally generates and emits RF energy. One common example is a remote garage door opener or Wi-Fi routers. These rarely cause interference to amateur radio unless they are operating within the amateur bands.
- Carrier current devices – A system that transmits RF energy by conduction through electric power lines.

FCC Part 18 Devices

- RF lighting devices, such as "grow" lights or other electronic ballasts, as well as lighting controllers.
- Induction cooking/ultrasonic equipment. These are rarely a cause of interference to amateur radio.

Important Rule: You don't need to know what it is in order to find it. Don't waste a lot of time analyzing the sound or trying to match it up with some other device. This is one of the biggest

---

mistakes that people make when confronted with an unknown source of radio interference from a consumer device.

Statistically, the most common problem reported to the American Radio Relay League (ARRL) is an unknown source of interference. Once the source is known, the most common reported problem is power line noise. After that, the most common sources are consumer devices, either in the complainant's home or a nearby residence.

Other Transmitters – There are several transmitter types that commonly cause RFI.

- Two-Way or Land Mobile Radio – Interference within a receiver passband can affect AM, FM or single-sideband (SSB) modulation. However, strong interfering FM signals may result in "capture effect", or overriding of the desired received signal.
- Paging Transmitters – Paging transmitters are generally very powerful FM or digital signals and will be obvious, if they fall in, or near, your receiver passband. Digital paging will sound very raspy, like a power saw or buzzing, and may interfere with a wide range of receive frequencies. Intermodulation distortion and adjacent channel overload caused by VHF

paging transmitters can be problematic in receivers tuned to the Amateur 2 meter band, as well as to VHF LMR. This intermodulation can occur at a transmitter site or can occur in the overload front end of an affected receiver. Fortunately, most of the VHF paging transmitters moved to the 929/931 MHz frequency pairs, so this is not the issue it once was.

- Broadcast Transmitters – Broadcast transmitter interference will have modulation characteristics similar to their broadcasts – AM, FM, video carriers, or digital signals. The video or digital signals will sound raspy or buzzing. Radio broadcasting also uses "remote pick-up" (RPU) using mobile communications vans in the 161, 450, and 455 MHz bands to link back to the studio. Some equipment used for these links can create RFI. Fortunately, this is a rare and generally a temporary issue.

Cable Television – Signal leakage from cable television systems will generally occur on their prescribed channel assignments. Many of these channels overlap existing over-the-air radio communications channels. For instance, leakage on cable TV analog channel 18 can cause interference to 2-meter amateur frequencies.

When the leaking signal is an analog TV channel, interference usually occurs on the visual carrier frequency, since that is where most of the TV channel's power is concentrated. An example in the 2-meter band would be cable TV channel 18's visual carrier on 145.25 MHz. If the leaking signal is a digital channel, interference will be similar to wideband noise (a digital cable channel is almost 6 MHz wide). One challenge with the latter is determining whether noise that appears to be leaking from a cable system is in fact a cable signal and not something else.

There have been several instances of wideband noise-like interference in the medium frequency (MF) and HF bands that was initially thought to be leakage of cable modem upstream digital signals. Further investigation found that the noise was typically from a Part 15 device or similar, and had nothing to do with cable system leakage. The noise was coupled to the cable TV lines via electrical code-required bonds between cable TV and telephone lines, and the power company neutral.

Wireless Network Interference – Interference to wireless networks (Wi-Fi, Bluetooth, etc.) is really outside the scope of this Guide, but you'll find some handy software tools in the References section for identifying interference and optimizing these networks. Real-time spectrum analyzers are also ideal for this type of interference identification.

Other Sources (ESD, Lightning, and Surge) – Not normally classified as "RFI", energy sources such as electrostatic discharge (ESD), nearby lightning and electrical power line surges can disrupt radio communications. Fortunately, these are brief events. ESD can change the state or mode of electronics or cause the processor to reset. Power line surges can cause damage to radio equipment, so external power line filtering and transient protection devices are recommended.

Sound Correlation for RFI Types

Demodulating Types of RFI – The recovered audio using the spectrum analyzer's demodulation/discriminator can also include tones, whistles, and more annoyingly, pops and clicks. This is also dependent upon the resolution bandwidth (RBW) settings that feed the demodulation/discriminator stage of the spectrum analyzer used. Of course, the demodulator may also be used to verify the identity of AM or FM transmitters.

Identifying Digital Communications Modes – For more information on identifying the sound of specific digital modes of transmission, check out the following web site: www.kb9ukd.com/digital/.

■ Locating RFI

As previously noted, statistically, the vast majority of interference complaints reported to the ARRL is caused either by RF emissions from consumer devices or power line noise. A small percentage of complaints are caused by other problems, such as intermodulation distortion or cable television leakage. In most cases, interference to amateur radio reception due to power line noise or nearby consumer devices is best located using a portable receiver, rather than a spectrum analyzer. Although it isn't initially necessary to know what the source is, knowing how to tell the difference between these broad categories can be helpful toward a timely solution.

Is It In Your Own Home or Nearby?

Consumer Devices – For locating consumer devices in your home, temporarily trip the main breaker to your residence while listening to the interference with a battery-powered radio. The noise will go away if the source is in your home.

In this case, you can then further isolate the source by tripping the individual circuit breakers, one at a time, until the noise goes away. Once you know the circuit, you can unplug the devices on that circuit one at a time to identify the noise source. Don't forget about the possible use of uninterruptible power supplies (UPS) or battery-powered devices that may continue to supply power to computer equipment or other equipment. You may need to disable these manually.

Faulty Light Switches or AC Sockets – One other potential source of intermittent arcing-type RFI could be worn out or failing light switches or arcing wires on AC outlet sockets.

If your home checks out OK, use a portable or mobile receiver and ensure the signal is received OK at your own residence. Check the signal strength at all nearby residences and adjust the sensitivity of the receiver down as you approach the potential source. A step attenuator inserted between the antenna and receiver is very handy in controlling receiver overload as the source is approached. Assuming the source device meets the applicable FCC emission limits, it will be located within a few hundred feet of your antenna. Frequently, it will also be located on the

---

same power transformer secondary system as the receiver exhibiting interference. Refer to ARRL RFI Book for more complete information on locating the source residence when the interference is caused by a consumer device.

Power Line Noise (PLN) – Information on how to handle a power line noise problem can be found on the Power Line Noise FAQ Page: www.arrl.org/power-line-noise-faq. If you're having trouble determining if a problem is PLN or a consumer device, see the question: "I'm having an RFI problem. How do I know if it's power-line noise and not some other electrical device?"

The best procedure for locating PLN sources is almost always to direction-find (DF) it. However, if the source is nearby, you may be able to identify the power pole with an alternative approach in some cases. Use a procedure similar to the one for locating a source residence when a consumer device causes the interference. The only difference is that you will be locating a utility pole as opposed to a residence. In either case, use a portable shortwave or higher-frequency radio, adjusted for AM reception and tuned off-station. Tune to the highest frequency you can and still hear the noise. A scanner-type radio with the AM aircraft band or a handheld VHF radio with AM

---

mode may also be used. If available, a directional antenna is best to pinpoint the power pole causing the interference; however, you'll most likely need an RF step attenuator or RF gain control.

Simple Direction Finding

DF Techniques – There are two primary methods for DFing. (1) "Pan 'N Scan" where you "pan" a directional antenna and "scan" for the interfering signal, recording the direction on a map, while keeping note of intersecting lines. (2) "Hot and Cold" where an omnidirectional antenna is used while watching the signal strength. In this method, the rule of thumb is for every 6 dB change you've either doubled or halved the distance to the interfering source. For example, if the signal strength was –30 dBm at one mile from the source, traveling to within a half-mile should read about –24 dBm on the spectrum analyzer. See the chart of dBm, etc., versus "S"–units for use with conventional communications receivers.

Note that when tracking PLN to a particular power pole, you will likely get several noise peaks, progressively getting stronger as you approach the noise source.

DF Systems – While radio direction-finding (RDFing) equipment can be installed into a

---

vehicle when attempting to locate distant sources, this is often not necessary if you don't mind the walk. There are several automated Doppler direction-finding systems available. Some recommended ones are:

- Antenna Authority (mobile, fixed and portable): www.antennaauthorityinc.com
- Doppler Systems (mobile and fixed): www.dopsys.com
- Rohde & Schwarz (mobile, fixed, and portable): http://www.rohde-schwarz.com

Step Attenuator – You'll also find a step attenuator quite valuable during the process of DFing. This allows control over the signal strength indication (and receiver overload) as you approach the interference source. Step attenuators may be purchased on sales sites, such as eBay, or through electronics distributors, such as DigiKey.

Locating Power Line Interference

For low-frequency interference – particularly power line noise – the interference path can include radiation due to conducted emissions along power lines. Therefore, when using the "Hot and Cold" method you'll need to be mindful that the radiated noise will generally follow the

---

route of the power lines, peaking and dipping along the route. The maximum peak usually indicates the actual noise source. As a complication, there may be several noise sources – some possibly very long distances away.

Use of VHF Receivers – Whenever possible, you'll generally want to use VHF or higher frequencies for RDFing. The shorter wavelengths not only help in pinpointing the source, they also make smaller handheld antennas more practical for RDFing.

A portable AM/shortwave receiver with telescoping antenna may also be used to successfully track down nearby consumer devices, especially if the noise does not affect VHF, thus facilitating an RDFing approach. Plans to build simple RDFing antennas for noise locating using low-cost "tape measure" elements are available online at www.arrl.org/files/file/Technology/HANDSON.pdf. The authors also recommend the commercially available "Arrow" antennas, or similar (see References section).

Signature Analysis – This is a powerful locating technique used by professional RFI locators. It is most useful for locating power line noise or consumer devices and the "electronic signature" it

---

records is very specific to a particular source. See the previously referenced Power Line Noise FAQ Page for additional details.

The typical use for signature analyzers is to locate arcing hardware on power poles. Once the source power line pole has been identified, the interfering source hardware on that pole can then often be identified from the ground using an ultrasonic pinpointer. See the May 1998 issue of *QST* for the article *A Home-made Ultrasonic Power Line Arc Detector* for plans to build such a pinpointer. This article is also available online at www.arrl.org/files/file/Technology/PLN/Ultrasonic_Pinpointer.pdf.

DFing HF Interference – While it can be a challenge, it is possible to locate sources of HF interference using portable radio receivers and directional antennas. Tom Thompson's (W0IVJ) article *Locating RF Interference at HF* (*QST*, November 2014), describes the method used. See Figure 4 for the general test setup. The article is linked on the ARRL RFI web site.

Rather than constructing your own antenna, a recommended commercial HF loop antenna would be the Scott Engineering LP-3, which is calibrated and is useful up to 15 MHz.

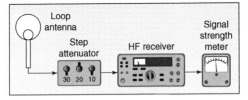

FIGURE 4 A block diagram of a simple high-frequency direction-finding setup (courtesy, ARRL).

The ARRL also has available the book, *Transmitter Hunting: Radio Direction Finding Simplified*, by Moell and Curlee. You'll find details in the References section.

Another good article, *Hunting Down RF Sources*, appeared in *QST*, February 2015 (page 45), and outlines an efficient process for locating sources of RFI in your own home. Refer to the ARRL RFI web site for a link.

Locating Narrow Band Interference

For most narrow band interference sources, such as co-channel, adjacent channel, and intermodulation interference, the recommended tool is the spectrum analyzer, as this allows you to "zero in" on particular frequency channels or bands and see "the big picture" of what's

---

occurring. Once the interfering signal is identified, the analyzer can then be used to DF the signal.

Using Spectrum Analyzers – Spectrum analyzers display frequency versus amplitude of RF signals. They can be helpful in determining the type and frequencies of interfering signals, especially for narrow band interference. There are two types of analyzers: swept-tuned and real time.

Swept-tuned analyzers are based on a superheterodyne principle using a tunable local oscillator and can display a desired bandwidth from start to stop frequencies. They are useful for displaying constant, or near constant, signals, but have trouble capturing brief intermittent signals, due to the lengthy sweep time.

A real-time analyzer samples a portion of the spectrum using digital signal processing techniques to analyze the captured spectrum. They are able to capture brief intermittent signals and are ideal for identifying and locating signals that may not even show up on swept analyzers. Most real-time bandwidths are limited to 27–500 MHz, maximum. The Signal Hound BB60C and Tektronix RSA306 are both relatively inexpensive real-time spectrum analyzers that are USB-powered and use a PC for control and display.

One important point to keep in mind regarding the use of spectrum analyzers is that because they have an untuned front end, they are particularly susceptible to high-powered nearby transmitters off frequency from where you may be looking. This can create internal intermodulation products (spurious responses) or erroneous amplitude measurements that are very misleading. When using spectrum analyzers in an "RF rich" environment, it's important to use bandpass filters or tuned cavities (duplexers, for example) at the frequency of interest.

Spectrum analyzers are also useful to characterize commercial broadcast, wireless, and land mobile communications systems. For wireless or intermittent interference, real-time analyzers work best. If used for tracking PLN, it's best to place the analyzer in "zero-span" mode to observe the amplitude variation. Placing the analyzer in "Line Sync" may also be helpful.

Be sure that whatever analyzer you use does not produce interfering signals in the frequency band of concern. This is especially true if using a USB-based analyzer with a portable laptop computer, as computers can generate strong signals that may mask or confuse the identification of the RFI.

Filtering

Consumer Products or Appliances – Many RFI issues may be reduced through installation of low-pass (or ferrite) filters on I/O (Input/Output) and power cables of interfering consumer products or I/O, audio, speaker and power line connections (Figure 5). Telephone lines may also be filtered using this technique.

Commercial AC line filters (Figure 6) may be installed on devices that may be introducing power line interference. If building your own, be

FIGURE 5 Larger ferrite toroid cores are useful for I/O or power line filtering.

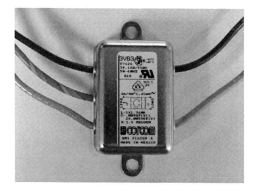

FIGURE 6 A typical AC line filter that may be added to equipment that is producing power line interference.

sure to use properly rated voltages – typically 1.5 kV rated capacitors.

Transmitters and Receivers – A spectrum analyzer may be used to check the harmonic content, frequency stability, and modulation quality of transmitters. Refer to the *ARRL RFI Book* for specific details.

One solution to RF coupling occurring through the microphone (or other I/O cables) is to simply use a clamp-on ferrite choke (Figure 7) on that cable, locating it close to the product enclosure.

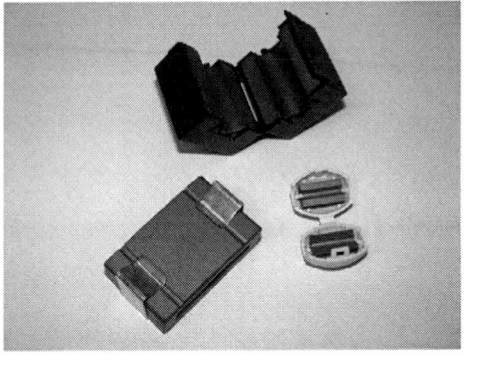

FIGURE 7 A sample of clamp-on ferrite chokes.

For HF applications, a toroid core may be required. Ferrite cores and chokes are available from Amidon, Fair-Rite, Laird, Würth Electronics, and several others (see References section).

If there's a strong adjacent signal interfering with reception (especially at transmitter sites), installing a bandpass filter tuned to the receive frequency may be the solution. For VHF or UHF frequencies, a surplus tuned cavity or duplexer tuned to the receiver frequency is a good solution.

Table of Ferrite Materials

Ferrite Type	Freq. Range (MHz)	Material	Suggested Usage
31	1–500	MnZn	EMI suppression
43	20–250	NiZn	EMI suppression
61	20–200	NiZn	EMI suppression and wideband transformers
64	Peak @ 400 MHz	NiZn	Best for VHF and UHF applications
73	Peak at 25 MHz	NiZn	Good general-purpose material for HF applications
75	Peak at 6 MHz	MnZn	AM broadcast interference and 160 meters

Information courtesy, Fair-Rite Products, Corp. (www.fair-rite.com)

Dealing With Neighboring Interference

When an interference source is located in a nearby residence, it is time to approach your neighbor. Diplomacy is of the utmost importance. You can print the following Neighbor pamphlet, www.arrl.org/information-for-the-neighbors-of-hams and the appropriate instructions (Breaker test on page 11.24 of *ARRL RFI Book*), for you as well as the neighbor. We generally recommend approaching them with a radio in hand, preferably an AM broadcast receiver, and have the noise

---

active when you knock on the door. Let them hear it but not so loud that it will be offensive. Inform them this is the problem you are experiencing and you believe the source may be in their home. Don't suggest what you think the cause is. If you're wrong, it often makes matters worse. Show them the info sheet and tell them it will only take a minute to eliminate their house.

Important Note: Recently, there has been a proliferation of illegal (and in some states, legal) indoor hydroponic marijuana "grow operations". These typically use high-powered lighting that utilizes non-compliant electronic ballasts (basically switched-mode power supplies) that create large amounts of interference. If you suspect this may be the case, it may be best to contact the FCC, rather than confront the owner.

Dealing With Intermodulation

To determine if you're dealing with an intermodulation product, perform an attenuator test as follows. Add a fixed amount of attenuation, such as 10 dB. If the signal drops by the amount of attenuation added, then it's not likely an intermodulation (IM) product. However, if it goes down by more than the expected amount, it may

---

be one. For example, adding 10 dB to the receiver may drop an intermodulation distortion (IMD) product by 30 dB. The fix in this case typically involves a filter on your radio, reducing the RF gain, or adding attenuation to the receiver.

For commercial communications applications, the addition of an Intermodulation Panel (IM Panel), comprised of an RF isolator/circulator sometimes with multiple stages followed with a low pass filter to knock down the harmonics of the circulator/isolator will resolve the problem. This is generally required at most commercial RF sites for any transmitter. Even amateur repeaters at these sites are not exempt from local site management policies on this. IM panels do work well and are recommended anywhere you have more than one transmitter, regardless of whether they are in different bands of the spectrum.

Dealing With Interference to Your Equipment

The usual issue of equipment susceptibility would be interfering signals to a receiver. Most interference will be single frequency signals from narrow band harmonics, receiver desensing from nearby transmitters, or broadband noise (power line noise, switch-mode power supplies, lighting devices, etc.), which might in some cases

---

manifest itself as an increase in the general noise floor of the receiver – especially at frequencies below 100 MHz.

However, one of the most common consumer devices that can cause RFI is a switching mode power supply. The interference in this case typically exhibits a regular and repeating pattern of peaks and nulls as you tune across the spectrum. The peaks are typically around 30–80 kHz apart, but wider spacings are possible. They often drift slightly over time.

The usual remedy would be to add filtering to I/O or power cables from the offending equipment or separation between the source and victim. In the case of power line noise, however, the only cure is often to correct the defect at the source. This is a job reserved for your utility company!

Getting Local Help (ARRL RFI technical committee)

Suggestions for forming a local committee are described in Chapter 18 of the *ARRL RFI Book*. If you want to see if there is any local help, contact your ARRL Section Manager (SM). SMs are listed on page 16 of *QST*. Your local ARRL affiliated ham club is another possible source for local help.

Filing Complaints With the FCC

Under federal law, the FCC has jurisdiction in matters involving interference to radio communications. Options on filing an interference complaint with the FCC include:

- By email by visiting The Consumer Help Center at http://consumercomplaints.fcc.gov. Information specific to an interference complaint, including a link for filing an online complaint, is also available at:

    https://consumercomplaints.fcc.gov/hc/en-us/articles/202916180-Interference-with-Radio-TV-and-Telephone-Signals

- By telephone by calling 1-888-CALL-FCC (1-888-225-5322). For RFI to public safety systems, the FCC is quite responsive.
- By mail at the following address. Be sure to include your name, address, contact information and as much complaint detail as possible:

    Federal Communications Commission
    Consumer and Governmental Affairs Bureau
    Consumer Inquiries and Complaints Division
    445 12th Street, S.W.
    Washington, DC 20554

Filing With ARRL

Most RFI and power line noise complaints involving Amateur radio are filed through the ARRL using the Cooperative Agreement process. See the Power Line Noise FAQ Page for complete details. The URL is www.arrl.org/power-line-noise-faq.

▨ Assembling an RFI Locating Kit

Eventually, you may wish to assemble a kit to locate, evaluate, and resolve RFI. Refer to the Manufacturers and Distributors reference near the end of this guide for web site links.

Portable Receivers – Low-cost AM or AM/SW broadcast receivers make good detectors for power line interference and consumer devices (Figure 8). The best ones will include signal strength indicators, wide intermediate frequency (IF) filters and RF gain controls. Start by tuning off-station in the AM mode. In general, you want to be using the highest frequencies at which you can hear the noise, or have a directional antenna. You'll generally be able to hear the noise, or "hash", at progressively higher frequencies as you approach the source. As you approach the source, you may wish to tune to higher frequencies in the shortwave or VHF/UHF bands in order to better pinpoint the source. Once you can hear the noise

---

FIGURE 8 The Grundig "Mini-400" AM/FM/SW receiver is just $40 (at the time of writing) and makes a low-cost power line noise detector.

in the Amateur 2 m band, you should be able to pinpoint the source pole with a 3 or 4 element handheld Yagi. The Tecsun PL-360 AM/FM/SW receiver is unique in that the signal strength meter is calibrated in dBuV (Figure 9).

A portable VHF receiver with AM aircraft band would also be useful for locating commercial power line noise from specific power poles. Most amateur VHF handheld transceivers have an AM mode, as well. If the receiver included an external antenna connection so a directional antenna could be connected, that would be a plus.

Spectrum Analyzers – Low-cost swept-tuned spectrum analyzers might include the RF Explorer (Figure 10), Rigol DSA815 (Figure 11),

---

FIGURE 9 Another good AM/FM/SW portable receiver is the Tecsun PL-360, available for about $50 (at the time of writing). This model includes a very sensitive "loopstick" directional AM antenna.

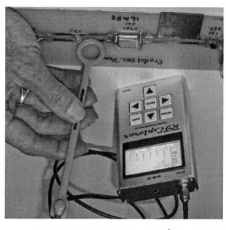

FIGURE 10 The RF Explorer WSUB3G spectrum analyzer covers 15 MHz to 2.7 GHz and costs just $269 (at the time of writing).

FIGURE 11 The Rigol DSA815 is an affordable spectrum analyzer that covers 9 kHz to 1.5 GHz. The cost is just $1,295 (at the time of writing). (Photo courtesy Rigol Electronics.)

or Thurlby Thandar PSA2702T (Figure 12). You may also find good used equipment buys on sites such as eBay or other used equipment dealers.

For intermittent interference (particularly for commercial communications installations), a real-time spectrum analyzer has the ability to capture these brief signals – some as brief as a few microseconds. Low-cost examples might include the Tektronix RSA306 (Figure 13) or Signal Hound BB60C (Figure 14).

For more serious interference location, Rohde & Schwarz has a portable system (Figure 15) that can quickly identify most interference sources and can also use the built-in mapping feature and GPS/compass in the antenna to triangulate the

---

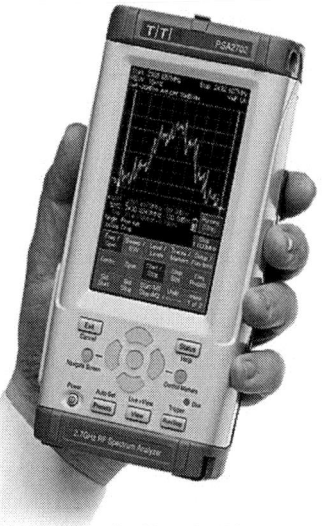

FIGURE 12 The Thurlby Thandar PSA2702T is an affordable portable spectrum analyzer that covers 1 MHz to 2.7 GHz. At the time of writing, the cost is just $1,695 from Saelig or Newark Electronics. (Photo courtesy Thurlby Thandar Instruments.)

source. Several fixed, mobile, or portable antennas are available for different frequency bands.

Narda has a similar interference analyzer, the Model IDA2 with a real-time bandwidth of 32 MHz and range of 9 kHz to 6 GHz.

FIGURE 13 The Tektronix RSA306 USB-controlled real-time spectrum analyzer covers 9 kHz to 6.2 GHz and has a real-time bandwidth of 40 MHz. The cost is $3,489 (at the time of writing).

FIGURE 14 The Signal Hound BB60C USB-controlled real-time spectrum analyzer covers 9 kHz to 6 GHz and has a real-time bandwidth of 27 MHz. The cost is $2,879 (at the time of writing). (Photo courtesy Signal Hound.)

FIGURE 15 The Rohde & Schwarz R&S®PR100 custom spectrum analyzer with mapping/triangulation and R&S®HE300 antenna. (Photo courtesy Rohde & Schwarz.)

Signature Analyzers – These are time-domain interference-locating instruments that produce a distinct "signature" of an interfering signal. This would include instruments produced by Radar Engineers (Figure 16). They are the best solution

FIGURE 16 A signature analyzer from Radar Engineers that tunes from 500 kHz to 1 GHz and which displays an electronic "signature" of a specific interference source. Receivers such as this are used by professional investigators to track down power line noise. (Photo courtesy Radar Engineers.)

for tracking down power line noise and consumer devices.

Antennas – For simply listening to power line noise, the built-in "loopstick" antenna on an AM broadcast band radio or telescoping antenna on a shortwave radio may work well. However, for tracking down power line noise to the source pole, and typically for RDFing other interfering sources, you'll want to use higher frequencies. A simple directional Yagi, such as the "measuring tape" antenna described previously or the Arrow II 146-4BP (Figure 17) with three-piece boom

---

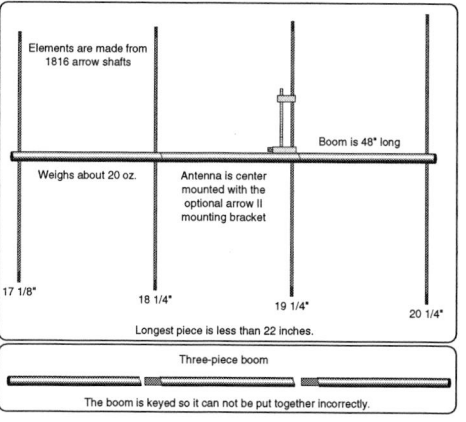

Arrow II back pack portable

Elements are made from 1816 arrow shafts

Boom is 48" long

Weighs about 20 oz.

Antenna is center mounted with the optional arrow II mounting bracket

17 1/8"

18 1/4"

19 1/4"

20 1/4"

Longest piece is less than 22 inches.

Three-piece boom

The boom is keyed so it can not be put together incorrectly.

FIGURE 17 Example of the Arrow Antenna Model Arrow II 146-4BP. (Photo courtesy Arrow Antennas.)

(www.arrowantennas.com) can be assembled quickly and attached to a short length of pipe and works well to receive RFI. In addition, the antennas may be dismantled and stored in a bag.

Also available are the low-cost (under $30 at the time of writing) PC board log-periodic antennas from Kent Electronics (www.wa5vjb.com) in Figure 18. These range in several frequency bands, starting at 400 MHz.

FIGURE 18 The larger PC board log-periodic antenna from Kent Electronics covers 400–1,000 MHz and costs under $30 (at the time of writing).

Step Attenuator – You'll also find a step attenuator quite valuable during the process of DFing. This allows control over the signal strength indication (and receiver overload) as you approach the interference source. The best are in steps of 10 dB and have a range of at least 80 dB, or more. Step attenuators may be purchased on sales sites, such as eBay, or through electronics distributors, such as DigiKey. Commercial sources would include Narda Microwave, Fairview Microwave, Arrow, and others. The ARRL also has do-it-yourself plans at this link:

www.arrl.org/files/file/Technology/tis/info/pdf/9506033.pdf.

Ferrite Cores and Chokes – RF currents on cables may usually (but not always) be reduced by clamping a ferrite choke around the cable nearest the source of RF. Adding a few of these chokes in various sizes would be helpful for troubleshooting. It's best to use a large (2.4 inch) toroid ferrite core of 31 or similar material with multiple turns through it for use in HF frequencies. This is a common cure for interference to (or from) consumer equipment. Beads and clamp-on ferrites are generally not effective at HF.

AC Line Filter – The Morgan Manufacturing 475-3 filter (www.morganmfg.us) is useful for troubleshooting a wide variety of conducted emission issues.

Miscellaneous – Adhesive copper tape is also useful for sealing enclosure joints temporarily during troubleshooting. Rolls of this tape may be purchased from electronics distributors at $30 (at the time of writing), or more, per roll. I've also found that "snail tape" (under $10, at time of writing) used in gardening may be substituted. This may be found in garden stores or on Amazon. Take care not to cut yourself on the sharp edges.

Aluminum foil is also handy as a troubleshooting tool for wrapping around an interfering product to assess whether additional shielding might help. Note that aluminum foil is not as effective at power line frequencies or for conducted emissions.

Finally, a selection of capacitors, resistors, inductors, and common-mode chokes is useful for applying filtering to I/O, microphone, and power line cables.

■ U.S. FCC Rules

Highlights of Part 15 (consumer devices)

The U.S. FCC specifies limits for product emissions, only. There are no immunity requirements. Part 15 addresses almost all electronic and non-electronic devices that can cause harmful interference. This includes both digital and non-digital devices.

There are four types of devices in Part 15: unintentional, incidental, intentional, and carrier current radiators. This would include information technology equipment (ITE) – for example, computers, printers, and associated equipment. It also includes most consumer products. There are a few exemptions, although most consumer devices that might cause an RFI problem are not exempt.

References include:

- § 15.103 Exempted Devices.
- § 15.109 Field strength limits. Describes unintentional radiators, i.e. most electronic consumer devices with circuitry operating at greater than 9 kHz.
- § 15.107 Conduction limits. Describes unintentional radiators, i.e. most electronic consumer devices with circuitry operating at greater than 9 kHz.
- § 15.19 Labelling requirements.
- § 15.21 Information to user.

Note: There are no specified emissions limits for incidental radiators under Part 15. However, as with all Part 15 and 18 consumer devices, operation of these devices must not cause harmful interference to a licensed radio service. If and when that happens, the burden to correct the problem then falls on the *operator* of the device.

Highlights of Part 18 (ISM and lighting)

ISM (Industrial, Scientific and Medical) devices convert RF directly into some other form of energy. For example, microwave ovens convert the RF into heat. Ultrasonic jewelry cleaners and humidifiers convert the RF into ultrasonic

---

(mechanical) energy. RF lighting devices convert the RF into light, and primary sources of reported RFI problems include electronic fluorescent light ballasts – especially large "grow lights" used for indoor hydroponic farms. Some of these have been measured to exceed the FCC emissions limits by a considerable margin. Compact fluorescent lamps (CFLs) are another Part 18 consumer device but ARRL hasn't received many complaints about them.

References include:

- § 18.305 Field strength limits. Describes some consumer devices, and paragraph (c) specifically talks about RF lighting devices.
- § 18.307 Conduction limits. Describes induction cooking ranges, Part 18 consumer devices, RF lighting devices, etc.
- § 18.213 Information to the user. Paragraph (d) refers to RF lighting devices.

▓ European Union (EU) Rules

The EMC Directive

All electronic products imported or manufactured in the European Union must meet the intention of the EMC Directive. Products that meet this are marked with a "CE" label. The EMC Directive merely states that a product must not interfere with the environment and the environment must not interfere with the product. This can be proven through testing to the appropriate EMC standards or through engineering analysis and recording the results into a technical construction file. Most other countries (worldwide) expect products to meet the same requirements. *Note*: amateur radio equipment and home-made electronic projects are exempt.

The EMC Directive includes both emissions and immunity requirements.

European Union (EU) Emission Limits

In general, the emission limits are more restrictive for residential environments and less restrictive for industrial environments. In the EU and the rest of the world, immunity, on the other hand, is usually more restrictive for industrial environments.

FCC/European Emission Limits

The limits for FCC (Part 15) and the European Union (CISPR) are very similar, but not identical.

■ Commonly Used Equations

Ohm's Law (formula wheel)

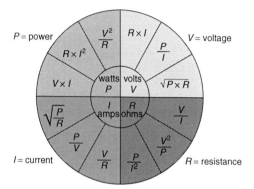

FIGURE 19 Ohms Law "formula wheel" for calculating resistance (R), voltage (V), current (I) or power (P), given at least two of the other values.

VSWR and Return Loss

VSWR (voltage standing wave ratio) given forward/reverse power

$$\text{VSWR} = \frac{1 + \sqrt{\dfrac{P_{\text{rev}}}{P_{\text{fwd}}}}}{1 - \sqrt{\dfrac{P_{\text{rev}}}{P_{\text{fwd}}}}}$$

VSWR given reflection coefficient

$$\text{VSWR} = \frac{1 + \rho}{1 - \rho}$$

Reflection coefficient, ρ, given the source and load impedances, Z_1, Z_2 in ohms

$$\rho = \left| \frac{Z_1 - Z_2}{Z_1 + Z_2} \right|$$

Reflection coefficient, ρ, given fwd/rev power

$$\rho = \sqrt{\frac{P_{\text{rev}}}{P_{\text{fwd}}}}$$

Return loss, given forward/reverse power

$$RL(\text{dB}) = 10 \log \left(\frac{P_{\text{fwd}}}{P_{\text{rev}}} \right)$$

Return loss, given VSWR

$$RL(\text{dB}) = -20 \log \left(\frac{\text{VSWR} - 1}{\text{VSWR} + 1} \right)$$

Return loss, given reflection coefficient

$$RL(\text{dB}) = -20 \log (\rho)$$

Mismatch loss given forward/reverse power

$$ML(\text{dB}) = 10\log\left(\frac{P_{\text{fwd}}}{P_{\text{fwd}} - P_{\text{rev}}}\right)$$

Mismatch loss, given reflection coefficient

$$ML(\text{dB}) = -10\log\left(1 - \rho^2\right)$$

E-Field from Differential-Mode Current

$$|E_{D,\max}| = 2.63 \times 10^{-14}\frac{|I_D|f^2 Ls}{d}$$

I_D = differential-mode current in a loop of area, A (m^2)

f = frequency (Hz)

L = length of loop (m)

s = spacing of loop (m)

d = measurement distance (3 m or 10 m, typ.)

(Assumption that the loop is electrically small and measured over a reflecting surface.)

E-Field from Common-Mode Current

$$|E_{C,\max}| = 1.257 \times 10^{-6}\frac{|I_C|f L}{d}$$

I_C = common-mode current in wire of length, L

f = frequency (Hz)

L = length of wire (m)

d = measurement distance (3 m or 10 m, typ.)

(Assumption that the wire is electrically short less than $1/2\lambda$.)

Antenna (Far Field) Relationships

Gain, dBi to numeric

$$\text{Gain}_{\text{numeric}} = 10^{(\text{dB}i/10)}$$

Gain, numeric to dBi

$$\text{dB}i = 10 \log (\text{Gain}_{\text{numeric}})$$

Gain, dBi to Antenna Factor

$$AF = 20 \log (\text{MHz}) - \text{dB}i - 29.79$$

Antenna factor to gain in dBi

$$\text{dB}i = 20 \log(\text{MHz}) - AF - 29.79$$

Field strength given watts, numeric gain, distance in meters

$$V/_m = \frac{\sqrt{30 \times \text{watts} \times \text{Gain}_{\text{numeric}}}}{\text{meters}}$$

Field strength given watts, dBi gain, distance in meters

$$V\!/_m = \frac{\sqrt{30 \times \text{watts} \times 10^{(\text{dBi}/10)}}}{\text{meters}}$$

Transmit power required, given desired V/m, antenna numeric gain, distance in meters

$$\text{watts} = \frac{(V/m \times \text{meters})^2}{30 \times \text{Gain}_{\text{numeric}}}$$

Transmit power required, given desired V/m, antenna dBi gain, distance in meters

$$\text{watts} = \frac{(V/m \times \text{meters})^2}{30 \times 10^{(\text{dBi}/10)}}$$

E-Field Levels Versus Transmitter Pout

Pout (W)	V/m at 1 m	V/m at 3 m	V/m at 10 m
1	5.5	1.8	0.6
5	12.3	4.1	1.2
10	17.4	5.8	1.7
25	27.5	9.2	2.8
50	38.9	13.0	3.9
100	55.0	18.3	5.5
1000	173.9	58.0	17.4

Assuming the antenna gain is numerically 1, or isotropic, and the measurement is in the far field and greater than 100 MHz.

Using Decibels (dB)

The decibel is always a ratio...

- Gain = P_{out}/P_{in}, where P = power
- Gain(dB) = $10 \log(P_{out}/P_{in})$, where P = power
- Gain(dB) = $20 \log(V_{out}/V_{in})$, where V = voltage
- Gain(dB) = $20 \log(I_{out}/I_{in})$, where I = current

(For the formulas for calculating gain in dB from voltages and currents, the input and output impedances must be identical.)

We commonly work with...

- dBc (referenced to the carrier level)
- dBm (referenced to 1 mW)
- dBμV (referenced to 1 μV)
- dBμA (referenced to 1 μA)

Power Ratios

3 dB = double (or half) the power

10 dB = $10 \times$ (or /10) the power

Voltage/Current Ratios

6 dB = double (or half) the voltage/current

20 dB − $10 \times$ (or /10) the voltage/current

dBm, dBμV, dBμA (conversion)

Log ↔ Linear Voltage

Volts to dBV	$dBV = 20 \log(V)$
Volts to dBμV	$dB\mu V = 20 \log(V) + 120$
dBV to Volts	$V = 10^{(dBV/20)}$
dBμV to Volts	$V = 10^{((dB\mu V - 120)/20)}$
dBV to dBμV	$dB\mu V = dBV + 120$
dBμV to dBV	$dBV = dB\mu V - 120$

Note: For current relationships, substitute A for V

In a 50 ohm system:

dBm to dBμV	$dB\mu V = dBm + 107$
dBm to dBμA	$dB\mu A = dBm + 73$
dBμV to dBμA	$dB\mu A = dB\mu V - 34$

Log Identities

If $Y = \log(X)$, then $X = 10^{Y}$

Log 1 = 0

Log of numbers > 1 are positive

Log of numbers < 1 are negative

The $\log (A \times B) = \log A + \log B$

The $\log (A/B) = \log A - \log B$

The log of $A^{n} = n \times \log A$

"S"-Units to dBm, dBµV, and µV (assuming a 50 Ω system) (*Note:* S9 = 50 µV and each S unit is 6 dB.)

"S"-Unit	dBm	dBµV(50 Ω)	µV(50 Ω)
S9 + 30 dB	−43	64	1,583
S9 + 20 dB	−53	54	500.5
S9 + 10 dB	−63	44	158.3
S9	−73	34	50
S8	−79	28	25
S7	−85	22	12.5
S6	−91	16	6.25
S5	−97	10	3.13
S4	−103	4	1.56
S3	−109	−2	0.78
S2	−115	−8	0.39
S1	−121	−14	0.20

Note that we're assuming that "S-9" is 50 µV, as defined by Collins Radio in the 1940s. Some receivers are calibrated differently. Unfortunately, few current amateur receivers follow this standard, so you may need to consider S-units as relative.

Note, also, that S1 (−121 dBm) is also the typical 12 dB SINAD point for narrow band FM (NBFM).

Commonly Used Power Ratios (dB)

Ratio	Power	Voltage or Current
0.1	–10 dB	–20 dB
0.2	–7.0 dB	–14.0 dB
0.3	–5.2 dB	–10.5 dB
0.5	–3.0 dB	–6.0 dB
1	0 dB	0 dB
2	3.0 dB	6.0 dB
3	4.8 dB	9.5 dB
5	7.0 dB	14.0 dB
7	8.5 dB	16.9 dB
8	9.0 dB	18.1 dB
9	9.5 dB	19.1 dB
10	10 dB	20 dB
20	13.0 dB	26.0 dB
30	14.8 dB	29.5 dB
50	17.0 dB	34.0 dB
100	20 dB	40 dB
1,000	30 dB	60 dB
1,000,000	60 dB	120 dB

Multiplying power by a factor of 2 corresponds to a 3 dB increase in power. This also corresponds to a 6 dB increase in voltage or current.

Multiplying power by a factor of 10 corresponds to a 10 dB increase in power. Multiplying a voltage or current by 10 is a 20 dB increase. Dividing by a factor of 10 corresponds to a 10 dB reduction in power, or 20 dB for voltage and current.

■ Useful Software

This software may be found by Internet-searching for the name or searching in iTunes or Google Play.

PC

Ekahau Wi-Fi Scanner (www.ekahau.com)

inSSIDer (www.inssider.com)

Kismet (www.kismetwireless.com)

Macintosh

Wi-Fi Explorer (www.adriangranados.com)

Wi-Fi Scanner (www.netspot.com)

iPad/iPhone

dB Calc (calculates/converts dB)

E Formulas (multitude of electronics-related formulas and calculators)

Interference Hunter from Rohde & Schwarz (includes a frequency look-up, harmonics calculator, wireless calculator, and zoomable U.S. spectrum allocation chart)

LineCalc (coax cable loss and electrical length)

μWave Calc from Keysight Technologies (μW/RF calculator)

RF Tools from Huber+Suhner (RF tools for reflection, frequency/wavelength, signal delay, impedance and dB)

RF Toolbox Pro (a comprehensive collection of RF-related tools and references)

Android

dB Calculator from Rohde & Schwarz (dB conversions)

Interference Hunter from Rohde & Schwarz (includes a frequency look-up, harmonics calculator, wireless calculator, and zoomable U.S. spectrum allocation chart)

RF & Microwave Toolbox (by Elektor, a collection of RF tools)

RF Engineering Tools (by Freescale, a compilation of calculators and converters for RF and microwave design)

RF Calculator (by Lighthorse Tech, wavelength, propagation velocity, etc.)

EMC & Radio Conversion Utility (by TRAC Global, converters, path loss, EIRP, wavelength, etc.)

▓ References

Books

ARRL, *The ARRL Handbook for Radio Communications*, 2015.

Gruber, Michael, *The RFI Book* (3rd edition), ARRL, 2010.

Loftness, Marv, *AC Power Interference Handbook* (2nd edition), Percival Publishing, 2001.

Moell, Joseph and Curlee, Thomas, *Transmitter Hunting: Radio Direction Finding Simplified*, TAB Books, 1987.

Nelson, William, *Interference Handbook*, Radio Publications, 1981.

Ott, Henry W., *Electromagnetic Compatibility Engineering*, John Wiley & Sons, 2009.

Witte, Robert, *Spectrum and Network Measurements* (2nd Edition), SciTech Publishing, 2014.

Magazines

InCompliance Magazine (www.incompliancemag.com)

Interference Technology (www.interferencetechnology.com)

QST (www.arrl.org)

Useful Web Sites

ARRL (www.arrl.org)

ARRL RFI Information
(http://www.arrl.org/radio-frequency-
interference-rfi)

FCC (http://www.fcc.gov)

FCC, Interference with Radio, TV and Telephone
Signals (http://www.fcc.gov/guides/interference-
defining-source)

IWCE Urgent Communications
(http://urgentcomm.com) has multiple articles on
RFI

Jackman, Robin, *Measure Interference in
Crowded Spectrum*, Microwaves & RF Magazine,
Sept. 2014 (http://mwrf.com/test-measurement-
analyzers/measure-interference-crowded-
spectrum)

Jim Brown has several very good articles on RFI,
including *A Ham's Guide to RFI, Ferrites,
Baluns, and Audio Interfacing* at:
(www.audiosystemsgroup.com)

RFI Services (Marv Loftness) has some good
information on RFI hunting techniques
(www.rfiservices.com)

Tektronix has a downloadable guide showing examples of various kinds of RF signals (http://info.tek.com/AM-RSA306-e-guide-to-RF-Signals.html)

TJ Nelson, Identifying Source of Radio Interference Around the Home, 10/2007 (http://randombio.com/interference.html)

Manufacturers and Distributors

Amidon (www.amidoncorp.com) – Ferrite and iron powder cores and toroids

Arrow Antennas (www.arrowantennas.com) – handheld Yagi antennas

Beehive Electronics (www.beehive-electronics.com) – near field probes

Fair-Rite (www.fair-rite.com) – ferrite chokes

Kent Electronics (www.wa5vjb.com) – PC board antennas

Morgan Manufacturing (www.morganmfg.us) – AC line filters

Narda Test Solutions (www.narda-sts.com) – DF systems

Radar Engineers (www.radarengineers.com) – signature analysis instruments for detecting RFI

Rigol Electronics (www.rigolna.com) – spectrum analyzer

Rohde & Schwarz (www.rohde-schwarz.com) – DF systems

Seeed Studio (www.seeedstudio.com) – RF Explorer spectrum analyzer

Signal Hound (www.signalhound.com) – real-time spectrum analyzer

Tektronix (www.tektronix.com) – real-time spectrum analyzer

Würth Electronics (www.we-online.com) – ferrite chokes

Standards Organizations

American National Standards Institute (www.ansi.org)

ANSI Accredited C63 (www.c63.org)

Society of Automotive Engineers (SAE, http://www.sae.org/servlets/works/committeeHome.do?comtID=TEVEES17)

Electromagnetic Compatibility Industry Association (UK, http://www.emcia.org)

CISPR (http://www.iec.ch/dyn/www/f?p=103:7:0::::FSP_ORG_ID,FSP_LANG_ID:1298,25)

IEC (http://www.iec.ch/index.htm)

Canadian Standards Association (CSA, www.csa.ca)

Federal Communications Commission (FCC, www.fcc.gov)

ISO (International Organization for Standardization) (http://www.iso.org/iso/home.html)

VCCI (Japan, Voluntary Control Council for Interference) (http://www.vcci.jp/vcci_e/)

IEEE Standards Association (www.standards.ieee.org)

SAE EMC Standards Committee (www.sae.org)

Common Symbols

A	Amperes, unit of electrical current
AM	Amplitude modulated
cm	Centimeter, one hundredth of a meter
dBc	dB below the carrier frequency
dBm	dB with reference to 1 mW
$dB\mu A$	dB with reference to 1 μA
$dB\mu V$	dB with reference to 1 μV
E	"E" is the electric field component of an electromagnetic field.
E/M	Ratio of the electric field (E) to the magnetic field (H), in the far-field; this is the characteristic impedance of free space, approximately 377 Ω

GHz	Gigahertz, one billion Hertz (1,000,000,000 Hertz)
H	"H" is the magnetic field component of an electromagnetic field.
Hz	Hertz, unit of measurement for frequency
I	Electric current
kHz	Kilohertz, one thousand Hertz (1,000 Hertz)
MHz	Megahertz, one million Hertz (1,000,000 Hertz)
m	Meter, the fundamental unit of length in the metric system
mil	Unit of length, one thousandth of an inch
mW	Milliwatt (0.001 watt)
mW/cm^2	Milliwatts per square centimeter (0.001 watt per square centimeter area), a unit for power density; 1 mW/cm^2 equals 10 W/m^2
P$_d$	Power density, unit of measurement of power per unit area (W/m^2 or mW/cm^2)
R	Resistance
V	Volts, unit of electric voltage potential
V/m	Volts per meter, unit of electric field strength
W/m^2	Watts per square meter, a unit for power density, 1 W/m^2 equals 0.1 mw/cm^2
λ	Lambda, symbol for wavelength; distance a wave travels during the time period necessary for one complete oscillation cycle
Ω	Ohms, unit of resistance

Ref: ANSI/IEEE 100-1984, *IEEE Standard Dictionary of Electrical and Electronics Terms*, 1984.

Acronyms and Definitions

AM (Amplitude Modulation) – A technique for putting information on a sinusoidal carrier signal by varying the amplitude of the carrier.

ARRL – American Radio Relay League, the U.S. national organization of amateur radio operators.

Audio Rectification – Semiconductor junctions can demodulate RF frequencies, which can disrupt analog circuits by changing voltage bias levels. This is especially the case where audio modulation is riding on the RF carrier frequency.

BDA (Bi-Directional Amplifier) – Used to improve mobile phone coverage inside large buildings.

Capture Effect – An issue with FM receivers, where a weaker received signal is totally blocked by a stronger signal.

CE (Conducted Emissions) – The RF energy generated by electronic equipment, which is conducted on power cables.

CI (Conducted Immunity) – A measure of the immunity to RF energy coupled onto cables and wires of an electronic product.

CISPR – French acronym for "Special International Committee on Radio Interference".

Conducted – Energy transmitted via cables or PC board connections.

Coupling Path – A structure or medium that transmits energy from a noise source to a victim circuit or system.

CS (Conducted Susceptibility) – RF energy or electrical noise coupled onto I/O cables and power wiring that can disrupt electronic equipment.

CW (Continuous Wave) – A sinusoidal waveform with a constant amplitude and frequency.

dBc – The dB value below the carrier amplitude. A reference to harmonics or spurious emissions in relation to the primary carrier frequency.

Demodulation – The process of separating the baseband information (audio, etc.) from the RF carrier.

DF (Direction Finding) – A technique used to locate the source of RF interference (RFI).

EMC (Electromagnetic Compatibility) – The ability of a product to coexist in its intended electromagnetic environment without causing or suffering disruption or damage.

EMI (Electromagnetic Interference) – When electromagnetic energy is transmitted from an electronic device to a victim circuit or system via

---

radiated or conducted paths (or both) and which causes circuit upset in the victim.

ESD (Electrostatic Discharge) – A sudden surge in current (positive or negative) due to an electric spark or secondary discharge causing circuit disruption or component damage. Typically characterized by rise times less than 1 ns and total pulse widths of the order of microseconds.

EU – European Union.

Far Field – When the distance from a radiating source increases far enough, the radiated field can be considered planar (or a plane wave). Most definitions assume this occurs at 1/6 th wavelength and from that point the E-field decreases as one over the distance.

FCC – U.S. Federal Communications Commission.

FM (Frequency Modulation) – A technique for putting information on a sinusoidal "carrier" signal by varying the frequency of the carrier.

HF – High frequency.

HVAC – Heating, ventilation and air-conditioning.

IEC – International Electrotechnical Commission.

IEEE – Institute of Electrical and Electronics Engineers.

IF – Intermediate frequency.

IM – Intermodulation.

IMD – Intermodulation distortion.

Industry Canada – The Canadian equivalent of the U.S. FCC.

Intermodulation Distortion (Transmitter, Receiver) – No amplifier is perfectly linear. As a result, signal mixing in an amplifier can generate false (typically undesired) signals in its output when two or more signals are present at the input. In a radio receiver, an intermodulation distortion (IMD) product can appear as an undesired "phantom" signal at one frequency when two or more strong stations appear at different frequencies in the input. IMD products generated within a transmitter can also appear as emissions outside its normal intended bandwidth. It is important to note that the IMD product frequencies are mathematically related to the signals that cause them. See Types of Interference in this Guide or Chapter 13 of the ARRL RFI Book, 3rd Edition for details.

I/O – Input/Output, usually referring to attached cables to a product

ISM (Industrial, Scientific and Medical equipment) – A class of electronic equipment including industrial controllers, test & measurement equipment, medical products and other scientific equipment. FCC Part 18 rules apply to ISM equipment, and some consumer devices also fall under Part 18. These include electronic fluorescent light ballasts and CFLs that operate at greater than 9 kHz, microwave ovens, and some ultrasonic jewelry cleaners. In each case, the RF is converted directly into some other form of energy.

ITE (Information Technology Equipment) – A class of electronic devices covering a broad range of equipment including computers, printers and external peripherals; also includes telecommunications equipment, and multimedia devices. FCC Part 15 rules apply to ISM equipment.

ITU – International Telecommunications Union.

IX – FCC designation for interference. See also, RFIX.

LED – Light-emitting diode.

LMR (Land Mobile Radio) – Mostly public service and commercial two-way radio systems.

MF – Medium frequency.

Near Field – When the distance to a radiating source is close enough that its field is considered spherical rather than planar. Typically considered about at less than 1/6 th of a wavelength from the source. Within the near field, the E-field generally decreases with distance squared and the H-field generally decreases with distance cubed.

Noise Source – A source that generates an electromagnetic perturbation or disruption to other circuits or systems.

NTIA (National Telecommunications and Information Administration) – This organization administers and manages the RF spectrum allocations for all U.S. federal agencies.

OFCOM – The British equivalent to the U.S. FCC.

PLN – Power line noise.

PLT (Power Line Transient) – A sudden positive or negative surge in the voltage on a power supply input (DC source or AC line).

Radiated – Energy transmitted through the air via antenna or loops.

RBW – Resolution bandwidth.

RDF (Radio Direction Finding) – See DF.

RE (Radiated Emissions) – The energy generated by a circuit or equipment, which is radiated directly from the circuits, chassis and/or cables of equipment.

RF (Radio Frequency) – A frequency at which electromagnetic radiation of energy is useful for communications.

RFI (Radio Frequency Interference) – The disruption of an electronic device or system due to electromagnetic emissions at radio frequencies (usually a few kHz to a few GHz). Also EMI.

RFIX (also RF IX) – FCC designation for RF interference.

RI (Radiated Immunity) –The ability of circuits or systems to be immune from radiated energy coupled to the chassis, circuit boards and/or cables. Also Radiated Susceptibility (RS).

RPU – Remote pick-up.

RS (Radiated Susceptibility) – The ability of equipment or circuits to withstand or reject nearby radiated RF sources. Also Radiated Immunity (RI).

Rusty Bolt Effect – Corrosion or rust built up between two pieces of metal create a semiconductor diode effect. Although this can result in external intermodulation products,

---

harmful interference caused by this phenomenon is rare and short range.

Spurious Emissions – Emissions usually caused by a transmitter due to non-linearities in the amplifier circuitry. This can also be caused when a transmitter is over-driven.

SSB – Single-sideband.

SW – Shortwave.

TVI (Television interference) – Interference to the video or audio of a television.

UHF – Ultra-high frequency.

UPS – Uninterruptible power supply.

VHF – Very high frequency.

Victim – An electronic device, component or system that receives an electromagnetic disturbance, which causes circuit upset.

VSWR (Voltage Standing Wave Ratio) – A measure of how well the load is impedance-matched to its transmission line. This is calculated by dividing the voltage at the peak of a standing wave by the voltage at the null in the standing wave. A good match is less than 1.2:1.

WISP (Wireless Internet Service provider) – These large Wi-Fi systems are installed in large

---

commercial buildings, such as hotels, corporations, or public areas.

XTALK (Crosstalk) – A measure of the electromagnetic coupling from one circuit to another. This is a common problem between one circuit trace and another.